Make it E

Age 8-9

Maths

Contents

Learning Activities

Quick Tests

Answers

Paul Broadbent and Peter Patilla

Numbers to *1000*

The numbers between 100 and 999 all have **three digits**.

$$376 \rightarrow 300 + 70 + 6$$

hundreds   tens   ones

When you add or subtract 1, 10 or 100, the digits change.

376 + 1 = 37**7**   376 + 10 = 3**8**6   376 + 100 = **4**76

I Continue these number chains.

a 757 → (+ 1) → 758 → (+ 1) → 759 → (+ 1) → 760

b 628 → (− 10) → 618 → (− 10) → 608 → (− 10) → 598

c 496 → (+ 10) → 506 → (+ 10) → 516 → (+ 10) → 526

d 641 → (+ 100) → 741 → (+ 100) → 841 → (+ 100) → 941

e 385 → (− 100) → 285 → (− 100) → 185 → (− 100) → 085

f 903 → (− 1) → 902 → (− 1) → 901 → (− 1) → 900

II Complete this number puzzle.

Across

1 Seven hundred and forty-three

5 Nine hundred and twenty

Down

2 Four hundred and nine

3 Three hundred and fifty-one

4 Six hundred and eight

Number sequences

A sequence is usually a list of **numbers in a pattern**.

Look at the difference between each number to spot the rule for the pattern.

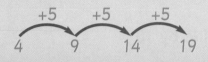

+5   +5   +5
4   9   14   19

The rule is +5

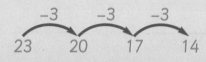

−3   −3   −3
23   20   17   14

The rule is −3

I Write the missing numbers in these sequences. What is the rule for each of them?

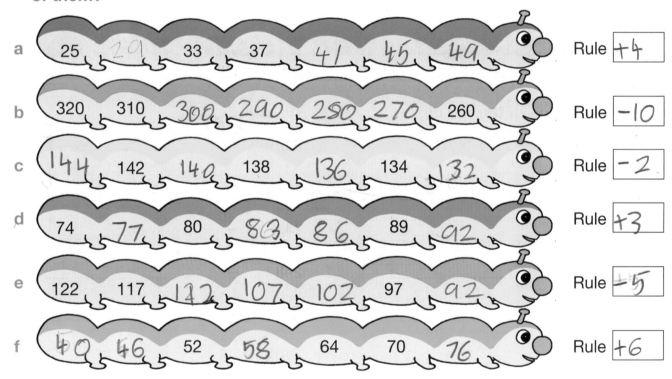

a 25 29 33 37 41 45 49   Rule +4

b 320 310 300 290 280 270 260   Rule −10

c 144 142 140 138 136 134 132   Rule −2

d 74 77 80 83 86 89 92   Rule +3

e 122 117 112 107 102 97 92   Rule −5

f 40 46 52 58 64 70 76   Rule +6

II Negative numbers go back past zero. Write the missing numbers on these number lines.

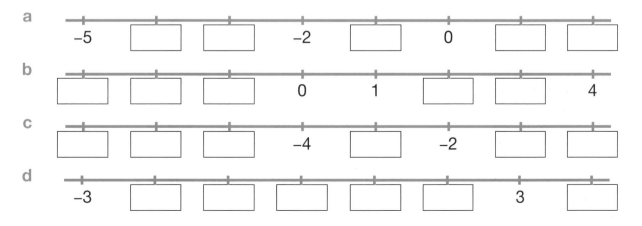

a −5 [ ] [ ] −2 [ ] 0 [ ] [ ]

b [ ] [ ] [ ] 0 1 [ ] [ ] 4

c [ ] [ ] −4 [ ] −2 [ ] [ ]

d −3 [ ] [ ] [ ] [ ] [ ] 3 [ ]

3

Place value

4-digit numbers are made from **thousands**, **hundreds**, **tens** and **ones**.

To make a number ten times bigger or ten times smaller you need to move the digits.

Multiply by 10

- Move the digits **one** place to the left.
- Fill the spaces with zero.

	3	8	2	× 10
3	8	2	0	

Divide by 10

- Move the digits **one** place to the right.

4	7	5	0	÷ 10
	4	7	5 . 0	

I Write the value of the red digit.

a 3450 → _5 tens_

b 6795 → _0 thousand_

c 4008 → _8 units_

d 9217 → _2 Hundreds_

e 3169 → _6 Tens_

f 5291 → _5 Thousands_

g 9469 → _9 units_

h 4778 → _7 hundreds_

i 7432 → _3 tens_

j 2984 → _4 units_

k 8898 → _8 hundreds_

l 4793 → _7 hundreds_

II Write the numbers coming out of each machine.

a 385   3850 ✓

b 790   7900 ✓

c 368   × 10   3680 ✓

d 412   4120 ✓

e 900   9000 ✓

f 4650   465 ✓

g 2910   291 ✓

h 3400   ÷ 10   340 ✓

i 5070   507 ✓

j 8000   800 ✓

Addition and subtraction

If you learn the addition and subtraction facts to 20, they can help you to learn other facts. Look at these patterns.

$4 + 9 = 13$

$40 + 90 = 130$

$400 + 900 = 1300$

$15 - 8 = 7$

$150 - 80 = 70$

$1500 - 800 = 700$

$5 + 7 = 12$

I Write the answers to these questions.

a
$7 + 5 = \boxed{12}$ ✓
$70 + 50 = \boxed{120}$ ✓
$700 + 500 = \boxed{1200}$ ✓

d
$13 - 6 = \boxed{9}$
$130 - 60 = \boxed{90}$
$1300 - 600 = \boxed{900}$

g $180 - 60 = \boxed{120}$ ✓

h $800 + 500 = \boxed{13000}$ ✓

i $1700 - 400 = \boxed{1300}$

b
$9 + 6 = \boxed{15}$
$90 + 60 = \boxed{150}$
$900 + 600 = \boxed{1500}$

e
$15 - 7 = \boxed{8}$
$150 - 70 = \boxed{50}$
$1500 - 700 = \boxed{500}$

j $1200 + 600 = \boxed{1800}$

k $150 + 90 = \boxed{240}$

l $130 - 90 = \boxed{140}$

c
$4 + 11 = \boxed{15}$
$40 + 110 = \boxed{150}$
$400 + 1100 = \boxed{1500}$

f
$18 - 9 = \boxed{9}$
$180 - 90 = \boxed{90}$
$1800 - 900 = \boxed{900}$

m $800 + 800 = \boxed{1600}$

II Circle touching pairs of numbers that total 100. The pairs can be vertical or horizontal. You should find ten pairs.

34	51	59	41	76	82	38	62
66	75	25	65	24	47	53	77
91	19	72	83	17	45	96	13
24	81	74	56	35	55	48	52

2-D shapes

A **polygon** is any 2-D shape with straight sides.

Count the number of sides to help name different polygons.

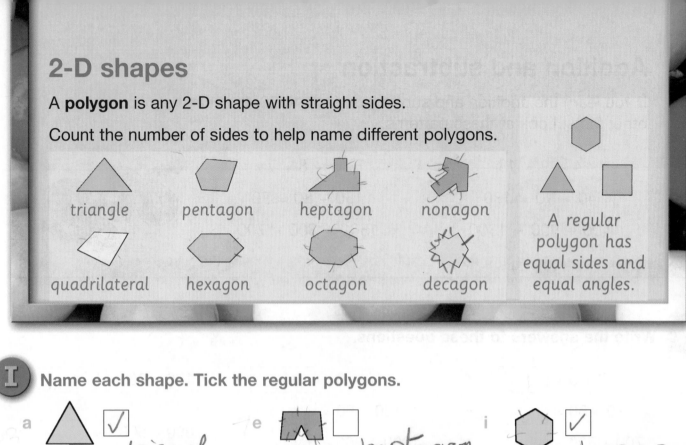

triangle   pentagon   heptagon   nonagon

quadrilateral   hexagon   octagon   decagon

A regular polygon has equal sides and equal angles.

I Name each shape. Tick the regular polygons.

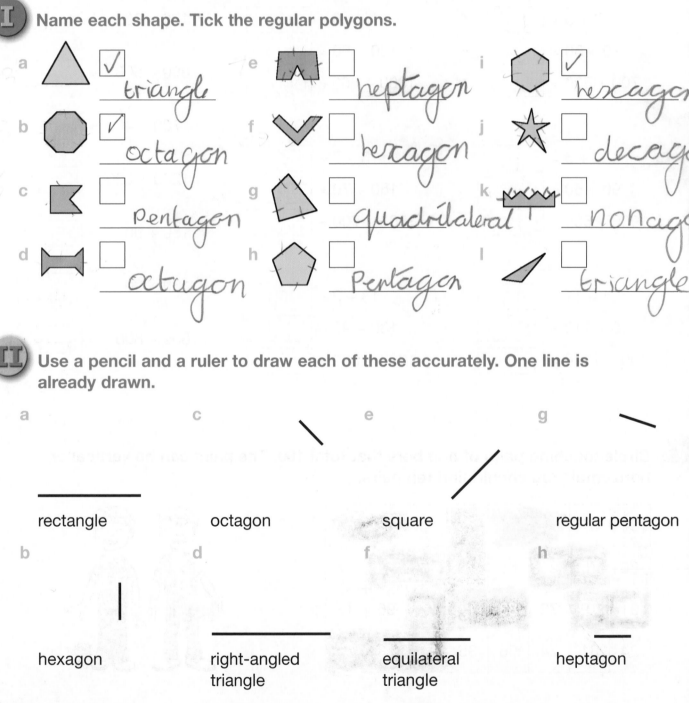

a ✓ triangle

b ✓ octagon

c ☐ Pentagon

d ☐ octugon

e ☐ heptagen

f ☐ hexagon

g ☐ quadrilateral

h ☐ Pentagon

i ✓ hexcagon

j ☐ decaga

k ☐ nonaga

l ☐ triangle

II Use a pencil and a ruler to draw each of these accurately. One line is already drawn.

a ____

rectangle

b |

hexagon

c octagon

d ____

right-angled triangle

e square

f ____

equilateral triangle

g square

regular pentagon

h

heptagon

Ordering numbers

To help work out the order of numbers, write them in a list. Make sure you line up the ones column.

| 7290 | 729 |
| 792 | 7209 |

Look at the numbers. Compare the **thousands**, then the **hundreds**, **tens** and finally the **ones** column.

| 729 |
| 792 |
| 7209 |
| 7290 |

I Write these in order, starting with the smallest.

a
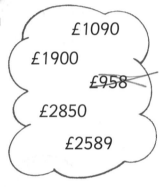
£1090
£1900
£958
£2850
£2589

b
3755 km
965 km
3095 km
3520 km
2830 km

c
2046 g
2460 g
2604 g
1599 g
1995 g

d
7025 ml
4599 ml
7529 ml
4600 ml
7028 ml

_____   _____   _____   _____
_____   _____   _____   _____
_____   _____   _____   _____
_____   _____   _____   _____
_____   _____   _____   _____

II Use the digits   .

Make as many different 4-digit numbers as you can. Write them in order, starting with the smallest.

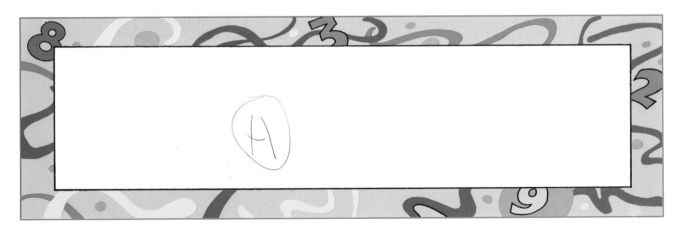

Time

On a clock face, read the **minutes past the hour** to tell the time.

As the minute hand moves around the clock, the hour hand moves towards the next hour.

5:42

42 minutes past 5

9:18

18 minutes past 9

I Write the times shown on each clock.

a

b

c

d

_____   _____   _____   _____

Draw the hands on these clocks.

e

7.56

f

9.03

g

3.41

h

11.18

II Write the number of minutes between each of these times.

a

[____] minutes

c

[____] minutes

b

[____] minutes

d

[____] minutes

Fractions of amounts

The number below the line of a fraction tells you how many parts to **divide** into.

$\frac{1}{4}$ of 8 is the same as $8 \div \mathbf{4} = 2$

$\frac{1}{3}$ of 15 is the same as $15 \div \mathbf{3} = 5$

I Use the pictures to help you answer these problems.

a
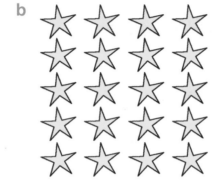

b

c

$\frac{1}{2}$ of 12 = 6 ✓

$\frac{1}{4}$ of 12 = 3 ✓

$\frac{1}{3}$ of 12 = 4 ✓

$\frac{1}{6}$ of 12 = 2 ✓

$\frac{1}{2}$ of 20 = 10 ✓

$\frac{1}{4}$ of 20 = 5 ✓

$\frac{1}{5}$ of 20 = 4 ✓

$\frac{1}{10}$ of 20 = 2 ✓

$\frac{1}{2}$ of 24 = 12 ✓

$\frac{1}{3}$ of 24 = 8 ✓

$\frac{1}{4}$ of 24 = 6 ✓

$\frac{1}{6}$ of 24 = 4 ✓

II Colour these grids to match the fractions. How many squares are left white on each? Make interesting patterns on each grid.

Blue

$\frac{1}{2}$ → red

$\frac{1}{4}$ → blue

$\frac{1}{6}$ → yellow

a
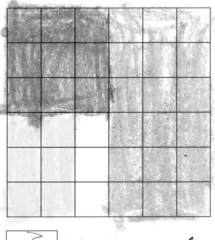

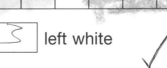
3 left white

b
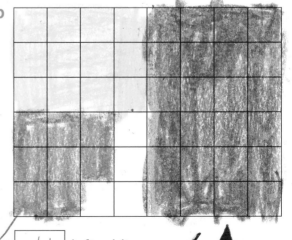
4 left white

✓ *Yellow*

✓ *Well done*

9

Measuring length

Look at these lengths.

10 millimetres (mm)	= 1 centimetre (cm)
100 cm	= 1 metre (m)
1000 m	= 1 kilometre (km)

Short lengths can be measured in millimetres.

Long distances can be measured in kilometres.

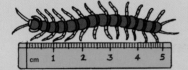

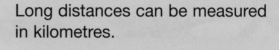

I Write these equivalent lengths.

a $3\frac{1}{2}$ km = ☐ m

b 40 mm = ☐ cm

c 150 cm = ☐ m

d 8 cm = ☐ mm

e $\frac{1}{4}$ m = ☐ cm

f 6500 m = ☐ km

g 22 cm = ☐ mm

h 18 km = ☐ m

i $4\frac{3}{4}$ m = ☐ cm

j 65 mm = ☐ cm

II Use a ruler to measure these lines in millimetres.

a ☐ mm

b ☐ mm

c ☐ mm

d ☐ mm

e ☐ mm

Multiplication and division

Multiplication and division are **linked**.

| 6 × 5 = 30 | If you know this, there are three other facts you also know. | 5 × 6 = 30
30 ÷ 5 = 6
30 ÷ 6 = 5 |

The three numbers 6, 5 and 30 are called a **trio**.

I Write four facts for each of these trios.

a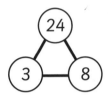

☐ × ☐ = ☐

☐ × ☐ = ☐

☐ ÷ ☐ = ☐

☐ ÷ ☐ = ☐

b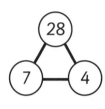

☐ × ☐ = ☐

☐ × ☐ = ☐

☐ ÷ ☐ = ☐

☐ ÷ ☐ = ☐

Write the missing numbers.

c ☐ × 6 = 36

d 8 × ☐ = 32

e ☐ ÷ 3 = 7

f 45 ÷ ☐ = 9

g 54 ÷ ☐ = 9

h ☐ ÷ 4 = 6

i 4 × ☐ = 16

j ☐ × 6 = 48

II If a number cannot be divided exactly, it leaves a remainder. Draw a line to join each division to its matching remainder.

60 ÷ 9

80 ÷ 3

37 ÷ 3

38 ÷ 6

93 ÷ 10

53 ÷ 6

106 ÷ 10

48 ÷ 5

89 ÷ 5

46 ÷ 6

61 ÷ 2

65 ÷ 6

Comparing numbers

The symbols > and < are used to compare numbers.

<	**>**
means 'is less than'	means 'is greater than'
729 < 750	2500 > 2100
729 is less than 750	2500 is greater than 2100

I Write the signs > or < for each pair of numbers.

a 455 ⊠ 396 ✓ g 3750 ⟩ 3079 ✓

b 817 ⟨ 870 ✓ h 6002 ⟨ 6010 ✓

c 958 ⟩ 936 ✓ i 5299 ⟨ 5300 ✓

d 1904 ⟨ 2301 ✓ j 7451 ⟩ 7415 ✓

e 1850 ⟩ 1508 ✓ k 5306 ⟨ 5311 ✓

f 2001 ⟩ 1998 ✓ l 9038 ⟩ 9009 ✓

II Write the numbers that could go in each middle box.

a 4169 > ☐ > 4164   4165 4166 4167 4168

b 3838 < ☐ < 3842

c 9002 > ☐ > 8996

d 4421 < ☐ < 4426

e 7082 > ☐ > 7076

3-D shapes

A **polyhedron** is a 3-D shape with flat faces.

A cube is a polyhedron. It has:

- 8 corners (vertices)
- 12 edges
- 6 faces.

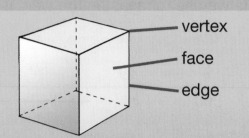

vertex

face

edge

I Name each shape. Choose the correct word from the box.

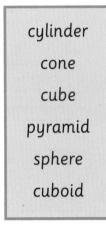

cylinder

cone

cube

pyramid

sphere

cuboid

a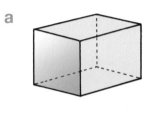

_____

c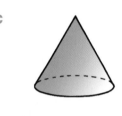

_____

e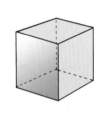

_____

b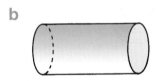

_____

d

_____

f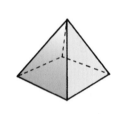

_____

II Write how many faces, edges and vertices each shape has.

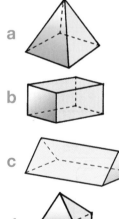

	faces	edges	vertices
a	_____	_____	_____
b	_____	_____	_____
c	_____	_____	_____
d	_____	_____	_____

13

Measuring mass

Kilograms (kg) and **grams** (g) are some of the units we use to measure the weight or mass of an object.

1000 g = 1 kg	250 g = $\frac{1}{4}$ kg
500 g = $\frac{1}{2}$ kg	750 g = $\frac{3}{4}$ kg

I Write these equivalent units.

a 2000 g = [ ] kg

b 1$\frac{1}{2}$ kg = [ ] g

c 5500 g = [ ] kg

d 1250 g = [ ] kg

e 7 kg = [ ] g

f 3$\frac{1}{4}$ kg = [ ] g

g 10 kg = [ ] g

h 6750 g = [ ] kg

i 2$\frac{1}{2}$ kg = [ ] g

j 4$\frac{3}{4}$ kg = [ ] g

k 9500 g = [ ] kg

l 1$\frac{3}{4}$ kg = [ ] g

II Look at these scales. Write the mass shown in kilograms.

a

_____

b
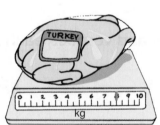

_____

c

_____

Write the mass shown in grams.

d

_____

e

_____

f
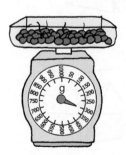

_____

Addition

When you add numbers, decide whether to use a **mental method**, or whether you need to use the **written method**.

Mental method 53 + 48

Example

> 53 add 50 is 103
> Take away 2 is 101

> 53 add 40 is 93
> 93 add 8 is 101

Written method 156 + 75

Example

```
  1 5 6
+   7 5
———————
  2 3 1
  1   1
```

Add the ones (6 + 5)

Then the tens
(50 + 70 + 10)

Then the hundreds
(100 + 100)

I Use your own methods to add these. Colour the star if you used a mental method.

a 51 + 43 = ☐ ☆   e 91 + 74 = ☐ ☆   i 88 + 83 = ☐ ✦

b 38 + 63 = ☐ ☆   f 57 + 69 = ☐ ☆   j 124 + 132 = ☐ ☆

c 29 + 35 = ☐ ☆   g 37 + 94 = ☐ ✦   k 146 + 105 = ☐ ☆

d 86 + 62 = ☐ ✦   h 75 + 66 = ☐ ☆   l 135 + 166 = ☐ ✦

II Answer these.

a
```
  3 8 6
+   5 8
———————

```

c
```
  5 4 6
+   7 4
———————

```

e
```
  6 7 6
+   7 8
———————

```

b
```
  2 7 4
+   8 1
———————

```

d
```
  9 1 4
+   8 7
———————

```

f
```
  7 2 7
+   8 3
———————

```

15

Area

The area of a shape can be found by **counting squares on a grid**.

Count half squares for shapes with straight sides.

For irregular shapes, count the squares that are covered more than half.

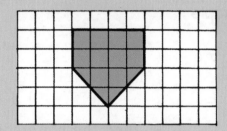

Area = 12 squares

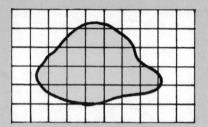

Area = 18 squares

I **Work out the area of each shape.**

Work out the approximate area of these shapes.

a Area =
☐ squares

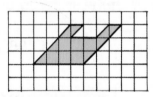

b Area =
☐ squares

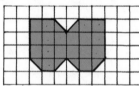

c Area =
☐ squares

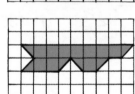

d Area =
☐ squares

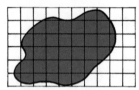

e Area =
☐ squares

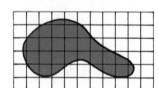

II **Draw three different shapes with an area of 8 squares.**

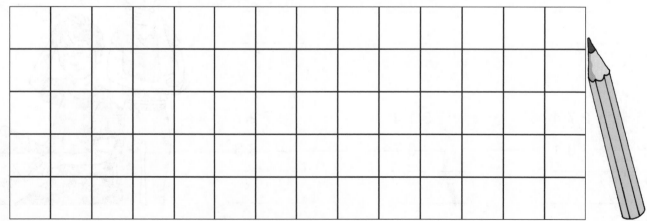

Measuring perimeter

The perimeter of a shape is the **distance all around the edge**.

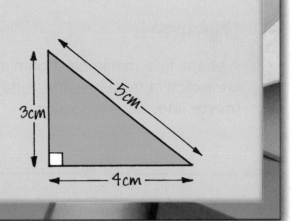

> The perimeter of this triangle is
> 3 cm + 4 cm + 5 cm = 12 cm

I Write the perimeter of each rectangle.

a Perimeter: ☐ cm

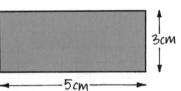
3cm
5cm

b Perimeter: ☐ cm

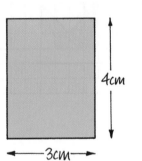
4cm
3cm

c Perimeter: ☐ cm

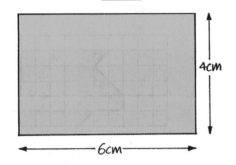
4cm
6cm

d Perimeter: ☐ cm

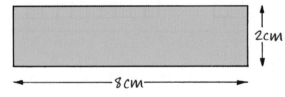
2cm
8cm

e Perimeter: ☐ cm

5cm
2cm

II Use a ruler to measure the perimeter of each shape.

a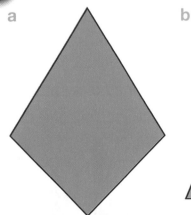

b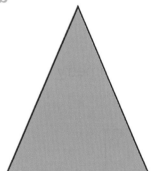

c

d

Perimeter =

_____

Perimeter =

_____

Perimeter =

_____

Perimeter =

_____

Symmetry

A shape is symmetrical if both sides are exactly the same either side of a **mirror line**, like a reflection.

This rectangle has two lines of symmetry.

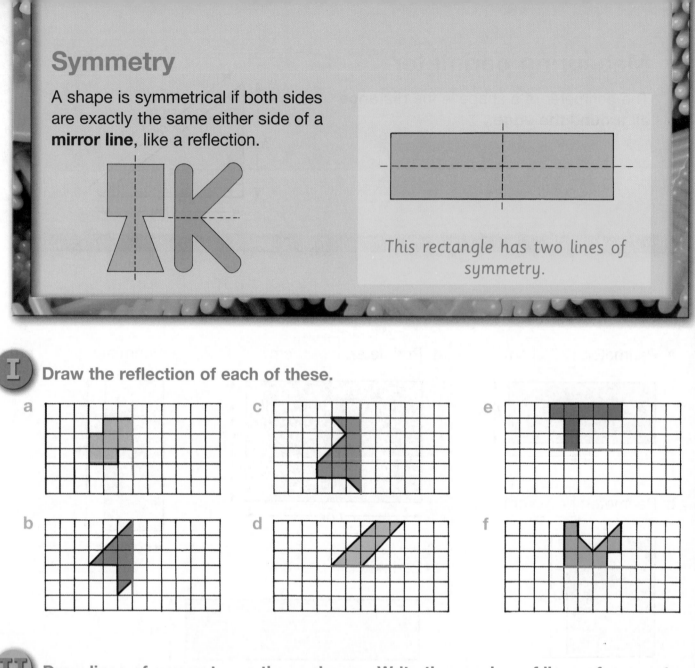

I Draw the reflection of each of these.

a   b   c   d   e   f

II Draw lines of symmetry on these shapes. Write the number of lines of symmetry for each shape.

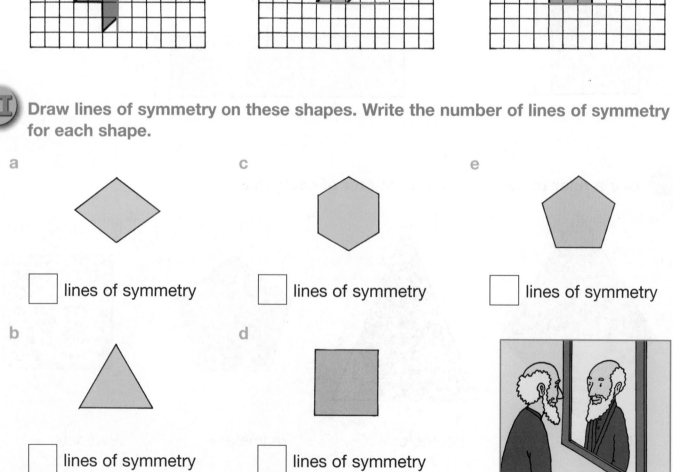

a   ⬜ lines of symmetry

c   ⬜ lines of symmetry

e   ⬜ lines of symmetry

b   ⬜ lines of symmetry

d   ⬜ lines of symmetry

Measuring capacity

Litres (l) and **millilitres** (ml) are some of the units we use to measure the capacity of liquids in containers.

1000 ml = 1 l	750 ml = $\frac{3}{4}$ l
250 ml = $\frac{1}{4}$ l	500 ml = $\frac{1}{2}$ l

I Write the equivalent units.

a 3000 ml = [   ] l

b 1$\frac{1}{2}$ l = [   ] ml

c 6 l = [   ] ml

d 2250 ml = [   ] l

e 1750 ml = [   ] l

f 10 l = [   ] ml

g 3$\frac{1}{2}$ l = [   ] ml

h 2000 ml = [   ] l

i 4500 ml = [   ] l

j 8$\frac{3}{4}$ l = [   ] ml

k 5750 ml = [   ] l

l 6$\frac{3}{4}$ l = [   ] ml

II Write the capacity each jug shows in millilitres.

a

[   ] ml

b

[   ] ml

c

[   ] ml

d

[   ] ml

e

[   ] ml

f

[   ] ml

g

[   ] ml

h

[   ] ml

Decimals

A **decimal point** is used to separate whole numbers from fractions.

$$0.1 = \tfrac{1}{10}$$

$$0.5 = \tfrac{1}{2}$$

$$1.2 = 1\tfrac{2}{10}$$

$$3.9 = 3\tfrac{9}{10}$$

$$45.6 = 40 + 5 + \tfrac{6}{10}$$

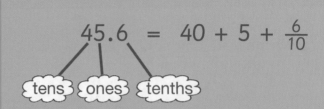

$$45.6 \; = \; 40 + 5 + \tfrac{6}{10}$$

tens   ones   tenths

I Change these fractions to decimals.

a $\tfrac{3}{10}$ =

b $\tfrac{1}{2}$ =

c $\tfrac{2}{10}$ =

d $\tfrac{7}{10}$ =

e $1\tfrac{7}{10}$ =

f $3\tfrac{3}{10}$ =

g $4\tfrac{1}{2}$ =

h $7\tfrac{9}{10}$ =

i $42\tfrac{1}{2}$ =

j $19\tfrac{4}{10}$ =

k $68\tfrac{8}{10}$ =

l $59\tfrac{6}{10}$ =

II Write the decimals on these number lines.

a

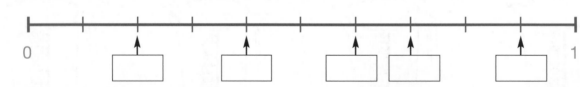

b

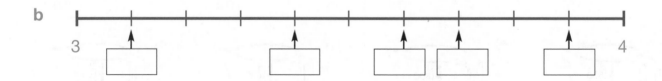

c

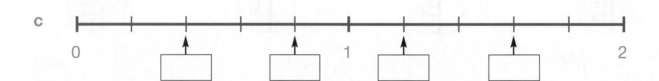

Reading pictograms

Pictograms are pictures or symbols to **show amounts**.

Check the number that each single picture represents.

This pictogram shows the number of skipping rope jumps in one minute by a group of children.

Jo	⋃ ⋃ ⋃ ⋃ ⋃ ⋃ ⸱
Sam	⋃ ⋃ ⋃ ⋃ ⋃
Alex	⋃ ⋃ ⋃ ⋃ ⋃ ⋃ ⋃ ⸱
Kim	⋃ ⋃ ⋃ ⋃
Ashley	⋃ ⋃ ⋃ ⋃ ⋃ ⋃ ⋃ ⋃ ⋃

⋃ = 5 jumps   ⸱ = 1, 2, 3 or 4 jumps

I Look at the skipping pictogram and answer these questions.

a How many skips did Sam jump? _____

b Who jumped the most in one minute? _____

c Who jumped a total of 37 skips? _____

d How many more skips did Ashley jump than Sam? _____

e Who jumped 12 fewer skips than Jo? _____

f If Jo jumped 32 skips, how many more is this than Sam? _____

g Write the number of skips for each child.

Jo: ☐   Sam: ☐   Alex: ☐   Kim: ☐   Ashley: ☐

II Carry out a 'skipping' survey. Ask family or friends to skip for a certain time and record the results as a pictogram. Decide on a symbol and a number this would represent.

name	number of skips

☐ = ☐ skips

Subtraction

When you subtract numbers, decide whether to use a **mental method**, or whether you need to use the **written method**.

Mental method 92 − 57 = 35

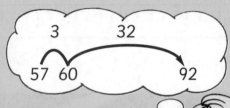

57 on to 60 is 3

60 on to 92 is 32

32 add 3 is 35

Written method 143 − 86 = 57

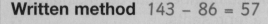
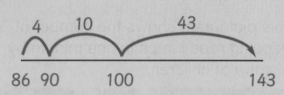

Count on from 86 in steps
4 + 10 + 43 = 57

These both use a number line to work out the answers.

I Use the number line method for these.

a 74 − 38 = ☐

38 ————————— 74

c 93 − 57 = ☐

57 ————————— 93

e 152 − 76 = ☐

76 ————————— 152

b 81 − 46 = ☐

46 ————————— 81

d 125 − 87 = ☐

87 ————————— 125

f 164 − 95 = ☐

95 ————————— 164

II Choose a method to work out the differences between these pairs of weights.

a
68kg 94kg

Difference: ☐ kg

c
91kg 31kg

Difference: ☐ kg

e
154kg 78kg

Difference: ☐ kg

b
84kg 56kg

Difference: ☐ kg

d

135kg 89kg

Difference: ☐ kg

f

171kg 94kg

Difference: ☐ kg

Equivalent fractions

Fractions that have the **same value** are called equivalent fractions.

$\frac{5}{10}$ is the same as $\frac{1}{2}$

$\frac{1}{3}$ is the same as $\frac{2}{6}$

I Complete the equivalent fractions.

a  $\frac{\square}{6} = \frac{\square}{3}$

e $\frac{1}{\square} = \frac{2}{8}$

h $\frac{2}{\square} = \frac{4}{10}$

b 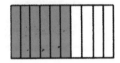 $\frac{\square}{10} = \frac{\square}{5}$

f $\frac{4}{12} = \frac{1}{\square}$

i $\frac{6}{\square} = \frac{1}{2}$

c $\frac{\square}{8} = \frac{\square}{2}$

g $\frac{\square}{10} = \frac{1}{2}$

j $\frac{\square}{4} = \frac{9}{12}$

d $\frac{\square}{8} = \frac{\square}{4}$

II Cross out the fraction that is not equivalent to the others in each set.

a $\frac{1}{2}$ →
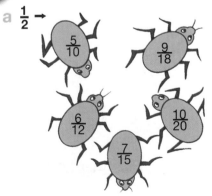

b $\frac{1}{4}$ →
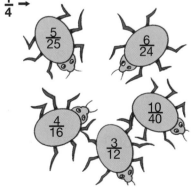

c $\frac{1}{3}$ →
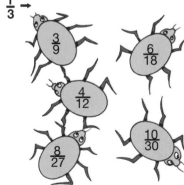

Rounding numbers

Rounding to the nearest 10

> 38 rounds **up** to 40
>
> 214 rounds **down** to 210

Look at the **ones** digit.

- If it is 5 or more, round up to the next tens.

- If it is less than 5, the tens digit stays the same.

Rounding to the nearest 100

> 653 rounds **up** to 700
>
> 439 rounds **down** to 400

Look at the **tens** digit.

- If it is 5 or more, round up to the next hundreds.

- If it is less than 5, the hundreds digit stays the same.

This chart shows a list of some of the highest waterfalls in the world. Round each height to the nearest 10 m and 100 m.

Waterfall	Country	Total drop (m)	Rounded to the: Nearest 10 m	Nearest 100 m
Angel	Venezuela	979	_____	_____
Tugela	South Africa	947	_____	_____
Mongefossen	Norway	774	_____	_____
Yosemite	USA	739	_____	_____
Tyssestrengane	Norway	646	_____	_____
Sutherland	New Zealand	581	_____	_____
Kjellfossen	Norway	561	_____	_____

Round these to the nearest 10 to work out approximate answers.

a 73 + 89 → ☐

b 346 − 152 → ☐

c 99 × 6 → ☐

d 814 + 338 → ☐

e 17 × 9 → ☐

f 509 − 296 → ☐

Multiples

Multiples are like the numbers in the **times tables**.

> Multiples of 2 are 2, 4, 6, 8, 10, 12 and so on.
>
> Multiples of 5 are 5, 10, 15, 20, 25 and so on.

Multiples of a number do not come to an end at ×10, they go on and on.
For example 52, 98, 114, 230 are all multiples of 2.

I Write these numbers in the correct boxes. Some of them will belong in more than one box.

48   56   100   39   86   52   82   42   63   85   70   115   60   65

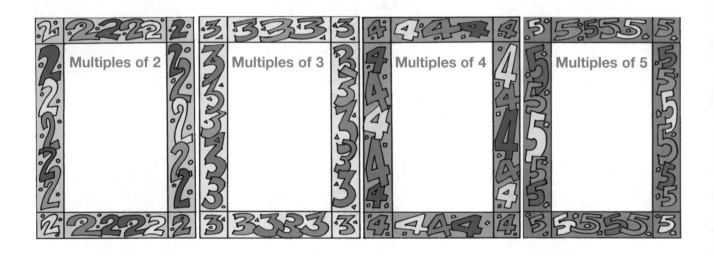

Multiples of 2   Multiples of 3   Multiples of 4   Multiples of 5

II Colour all the multiples of 3 red.
Colour all the multiples of 5 **blue**.
Look at the patterns on the grid.

1	2	3	4	5	6	7	8	9	10
11	12	13	14	15	16	17	18	19	20
21	22	23	24	25	26	27	28	29	30
31	32	33	34	35	36	37	38	39	40
41	42	43	44	45	46	47	48	49	50
51	52	53	54	55	56	57	58	59	60
61	62	63	64	65	66	67	68	69	70
71	72	73	74	75	76	77	78	79	80
81	82	83	84	85	86	87	88	89	90
91	92	93	94	95	96	97	98	99	100

Money problems

If you need to find the **difference** between two amounts, count on from the lower amount.

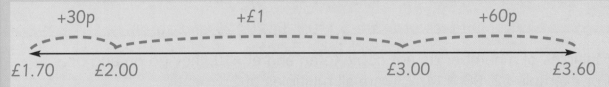

+30p   +£1   +60p

£1.70   £2.00   £3.00   £3.60

The difference between £1.70 and £3.60 is £1.90 (30p + £1 + 60p)

You can work out an amount of change in this way as well.

I Work out these differences.

a

£3.45
£1.70

Difference [ ]

c

£1.96   £3.40

Difference [ ]

e

£1.79
£2.07

Difference [ ]

b

Ready Meal
£1.40
£2.75 Fruit Cake

Difference [ ]

d

£3.05   £1.60

Difference [ ]

f

£2.84
£3.60

Difference [ ]

II Draw a line to join these price labels to the correct change from £10.

a £3.49

b £8.99

c £7.89

d £3.69

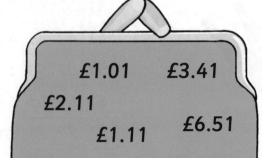

£1.01   £3.41
£2.11
£1.11   £6.51
£2.41   £2.61
£6.31

e £7.59

f £8.89

g £7.39

h £6.59

Angles

Angles are measured in degrees (°).

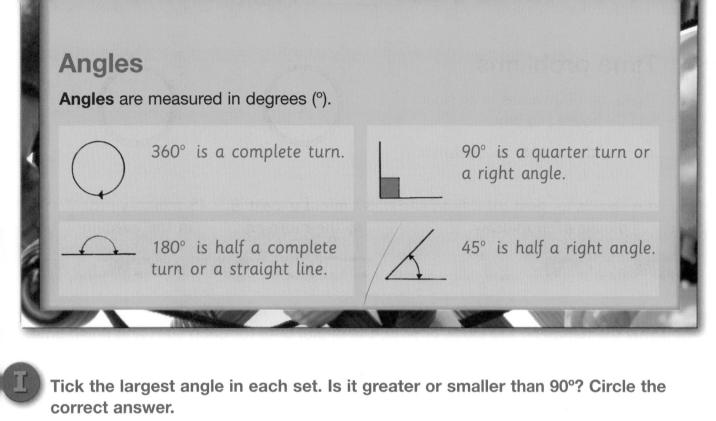

360° is a complete turn.	90° is a quarter turn or a right angle.
180° is half a complete turn or a straight line.	45° is half a right angle.

I Tick the largest angle in each set. Is it greater or smaller than 90°? Circle the correct answer.

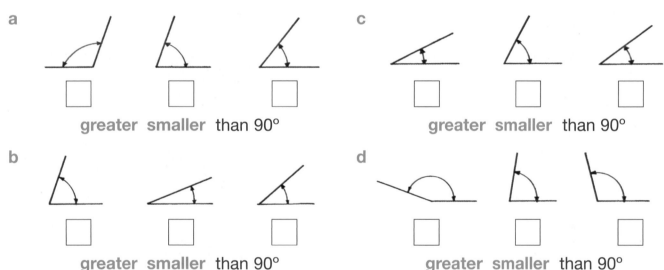

a

□ □ □

greater smaller than 90°

c

□ □ □

greater smaller than 90°

b

□ □ □

greater smaller than 90°

d

□ □ □

greater smaller than 90°

II These are the eight compass directions. Write the direction you will face after turning.

a Start facing north. Turn 90° clockwise. _____

b Start facing west. Turn 180° anticlockwise. _____

c Start facing south. Turn 45° clockwise. _____

d Start facing east. Turn 360° anticlockwise. _____

e Start facing north-east. Turn 90° clockwise. _____

f Start facing north-west. Turn 45° anti-clockwise. _____

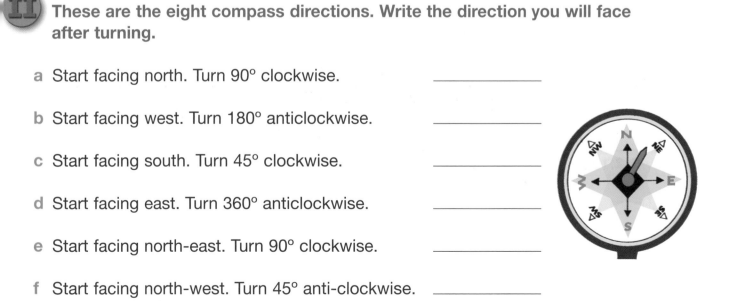

Time problems

There are 60 minutes in an hour and 24 hours in a day.

am stands for **ante meridiem** and means **before midday**.

pm stands for **post meridiem** and means **after midday**.

5.35am

35 minutes past 5 in the morning

7.15pm

15 minutes past 7 in the evening

Draw the hands on the clock or write the digital time for each start and finish time.

| Start | | Finish |

a  Mark goes swimming at 10.15am. He gets home 1½ hours later.

b  A train leaves London at 6.20pm. It arrives at Leeds 2 hours 20 minutes later.

c  Becky goes shopping at 11.10am. She finishes 3 hours 45 minutes later.

d 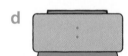 A football match starts at 1.45pm. It finishes 90 minutes later.

This timetable shows the times of buses. If you are at a bus stop at these times, how long will you have to wait?

BUS TIMETABLE

7.40am	8.15am	9.20am	10.50am	11.40am
2.10pm	4.30pm	5.10pm	6.30pm	8.00pm

a 9.05am   = _____ minutes   d 10.35am   = _____ minutes

b 11.15am   = _____ minutes   e 7.40pm   = _____ minutes

c 5.05pm   = _____ minutes   f 4.45pm   = _____ minutes

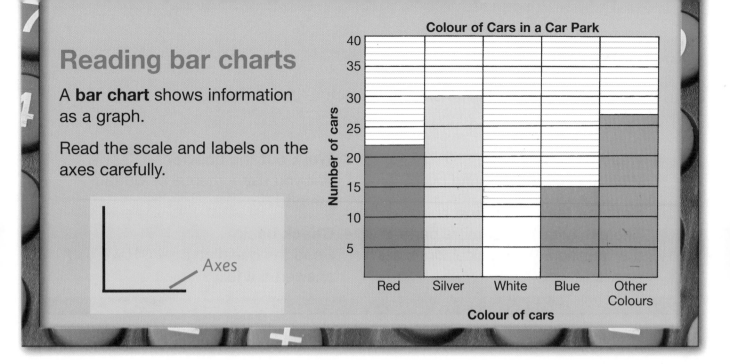

Reading bar charts

A **bar chart** shows information as a graph.

Read the scale and labels on the axes carefully.

Axes

Colour of Cars in a Car Park

Number of cars

Red   Silver   White   Blue   Other Colours

Colour of cars

 I Look at the graph above and answer these.

a Which colour was the most common car colour in the car park? _____

b How many cars were white? _____

c How many more cars were red than white? _____

d Which colour had half the number of silver cars? _____

e Black was the most common 'Other colour', with $\frac{1}{3}$ of these cars black. How many cars in total were black? _____

f How many cars in total were in the car park? _____

 II This graph shows the number of cars visiting a car wash over five days.

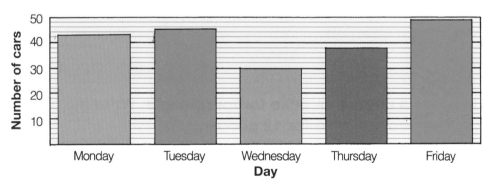

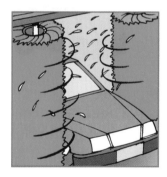

Number of cars

Monday   Tuesday   Wednesday   Thursday   Friday

Day

a How many cars visited the car wash on Tuesday? _____

b On which day did 38 cars visit the car wash? _____

c How many more cars visited on Friday than Monday? _____

d On which day did 15 fewer cars visit the car wash than on Tuesday? _____

Problems

When you read a **word problem**, try to 'picture' the problem.

Try these four steps.

1 Read the problem
What do you need to find out?

2 Sort out the calculation
There may be one or more parts to the question. What calculations are needed?

3 Work out the answer
Will you use a mental or written method?

4 Check back
Read the question again. Have you answered it fully?

I Read these word problems and answer them.

a A bar of chocolate costs 45p.
What do 4 bars cost? _____

b Sophie has 90 g of butter. She uses 35 g to make a loaf of bread.
How much butter is left? _____

c A board game costs £8.40.
It is reduced by £2.50 in a sale.
What is the new price of the game?

_____

d 68 people are going on a trip.
Minibuses can take 10 people.
How many minibuses will be needed?

_____

e A pencil costs 19p. How many can be bought for £2? _____

f Sam has £39 to spend on cinema tickets. If cinema tickets cost £4, how many tickets can he buy?

_____

g Mrs Benson travels 48 km each day to get to work and back. How far will she travel in 5 days?

_____

h 73 sheets of paper are put into folders that each hold 5 sheets of paper. How many folders are needed?

_____

II These are the ingredients of a chocolate cake for four people. Write the ingredients needed for a chocolate cake for 12 people.

50 g margarine
40 g sugar
60 g flour
1 egg
15 g cocoa powder
20 ml milk

_____
_____
_____
_____
_____
_____

Coordinates

Coordinates help to **find a position** on a grid.

Look at the coordinates of A and B.

Read the numbers across **horizontally** and then up **vertically** for the pair of coordinates. (2,6) and (7,3)

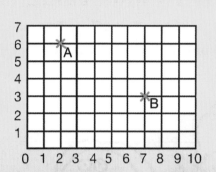

 Look at the grid below and answer the questions.

a What letter is at position:

(2,3) ____ (8,2) ____ (10,9) ____

b What are the coordinates for

D (___,___)   A (___,___)   S (___,___)

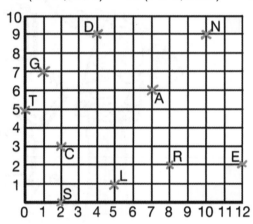

c Use the coordinates to spell out a shape and draw it in the box below.

(8,2) (12,2) (2,3) (0,5) (7,6) (10,9) (1,7) (5,1) (12,2)

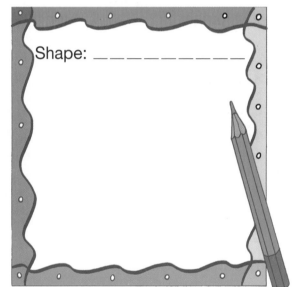

Shape: _____

 Draw a quadrilateral on this grid.

The coordinates are:

(4,2) (6,5) (3,7) (1,4)

The shape is a _____ .

Move two of the coordinates to make the shape into a rectangle.

Write the coordinates of your rectangle.

_____ _____

_____ _____

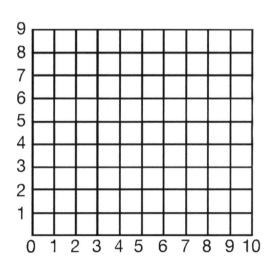

Test 1 Place value (1)

Thousands	Hundreds	Tens	Units	
4	9	5	7	= 4000 + 900 + 50 + 7

Write the missing numbers.

1. 4173 = 4000 + 100 + ☐ + 3

2. 8465 = ☐ + 400 + 60 + 5

3. 3657 = 3000 + ☐ + 50 + 7

4. 7895 = 7000 + 800 + ☐ + 5

5. 6218 = ☐ + 200 + 10 + 8

Write these as numbers.

6. two thousand one hundred and eight ☐

7. four thousand and ninety ☐

8. seven thousand two hundred and thirty-five ☐

9. three thousand eight hundred and sixteen ☐

10. nine thousand seven hundred ☐

Colour in your score

32

Test 2 Addition and subtraction

Knowing **number facts** can help you to work out other calculations.

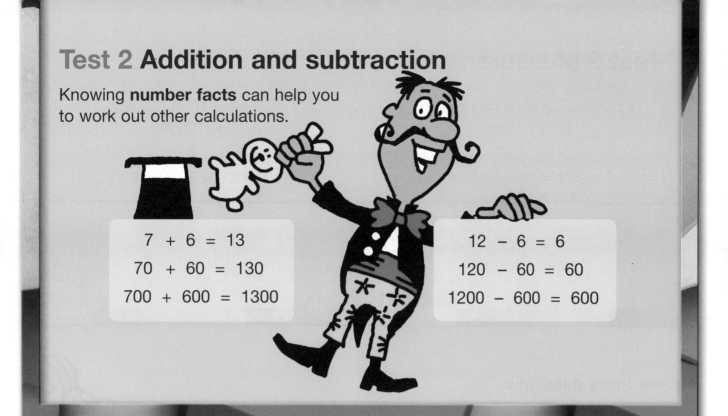

7 + 6 = 13
70 + 60 = 130
700 + 600 = 1300

12 − 6 = 6
120 − 60 = 60
1200 − 600 = 600

Answer these.

1.   40   +   70   =

2.   90   −   30   =

3.   130   −   50   =

4.   600   +   800   =

5.   900   −   400   =

6.   1200   +   500   =

7.   900   +   700   =

8.   170   −   80   =

9.   190   −   120   =

10.   800   +   500   =

Colour in your score

33

Test 3 **Measures**

1 centimetre = 10 millimetres
1cm = 10mm

1 litre = 1000 millilitres
1l = 1000ml

1 metre = 100 centimetres
1m = 100cm

1 kilogram = 1000 grams
1kg = 1000g

1 kilometre = 1000 metres
1km = 1000m

Answer these questions.

1. $\frac{1}{2}$ m = ⬚ cm

4. $\frac{1}{4}$ l = ⬚ ml

2. $\frac{1}{2}$ cm = ⬚ mm

5. $\frac{1}{4}$ kg = ⬚ g

3. $\frac{1}{10}$ km = ⬚ m

Measure these lines with a ruler.

6.  ⬚ mm

7. _____ ⬚ mm

8. _____ ⬚ mm

9.  ⬚ mm

10. _____ ⬚ mm

Colour in your score

34

Test 4 **2-D shapes**

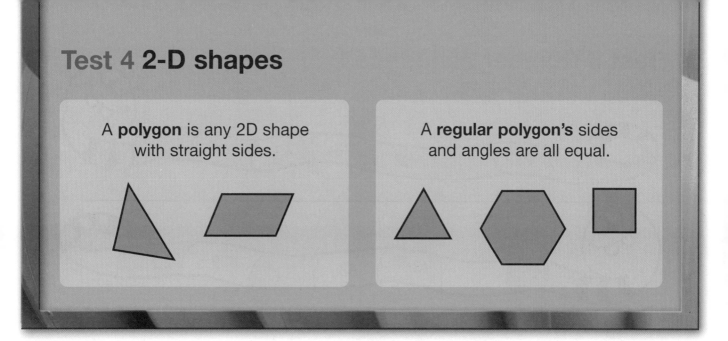

A **polygon** is any 2D shape with straight sides.

A **regular polygon's** sides and angles are all equal.

How many sides have each of these shapes?

1. A quadrilateral has [ ] sides.

2. An octagon has [ ] sides.

3. A hexagon has [ ] sides.

4. A triangle has [ ] sides.

5. A pentagon has [ ] sides.

Name these shapes.

6. _____

7. _____

8.  _____

9.  _____

10. _____

Colour in your score

Test 5 Number sequences

Number patterns can go up or down.

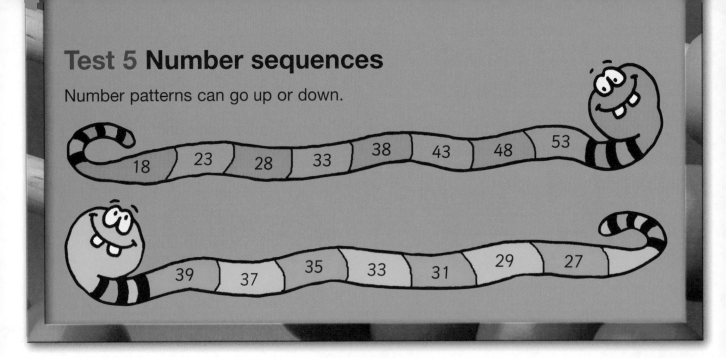

Write the missing numbers in these sequences.

1. | 32 | 35 | 38 | | 44 | 47 | 50 | 53 | | 59 |

2. | 48 | 52 | | 60 | 64 | 68 | | 76 | 80 | 84 |

3. | 31 | 29 | 27 | | 23 | 21 | | 17 | 15 |

4. | 230 | 210 | | 170 | 150 | | 110 | 90 | 70 |

5. | 76 | 81 | 86 | 91 | | 101 | 106 | | 116 |

Write the missing numbers on these number lines.

6. 7.

– 5 – 4 – 3 [ ] – 1 0 1 2 3 [ ] 5

8. 9. 10.

[ ] – 6 – 5 – 4 [ ] – 2 [ ] 0 1 2 3

10
9
8
7
6
5
4
3
2
1

Colour in your score

36

Test 6 **Multiplication tables**

You need to know your **tables**.

Remember, **4 x 6** is the same as **6 x 4**.

It doesn't matter which way round you multiply.

Write the missing numbers.

1.   7 × ☐ = 35

2.   ☐ × 4 = 40

3.   8 × 3 = ☐

4.   ☐ × 7 = 28

5.   2 × ☐ = 18

6.   8 × ☐ = 40

7.   10 × 6 = ☐

8.   ☐ × 3 = 27

9.   ☐ × 6 = 36

10.   4 × ☐ = 36

Colour in your score

Test 7 **Money**

£1 = 100p

£1.50 = 150p

£0.75 = 75p

£3.25 = 325p

Convert these amounts into pounds or pence.

1. £2.35 = ☐ p   4. £1.09 = ☐ p

2. £6.45 = ☐ p   5. £ ☐ = 214p

3. £ ☐ = 370p   6. £2.75 = ☐ p

Write the totals.

7. £1.85 • 70p • £ _____

8. 65p • £2.50 • £ _____

9. 90p • £3.15 • £ _____

10. £1.90 • £2.20 • £ _____

Colour in your score

38

Test 8 Fractions (1)

Fractions which are the same value are called **equivalent fractions**.

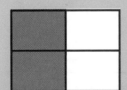 

$\frac{2}{4}$ is the same as $\frac{1}{2}$

 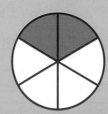

$\frac{1}{3}$ is the same as $\frac{2}{6}$

Write the fractions which are shaded.

1.  $\quad \frac{\boxed{}}{10} = \frac{\boxed{}}{5}$

2. $\quad \frac{\boxed{}}{6} = \frac{\boxed{}}{2}$

3. 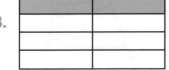 $\quad \frac{\boxed{}}{8} = \frac{\boxed{}}{4}$

4. 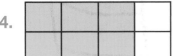 $\quad \frac{\boxed{}}{8} = \frac{\boxed{}}{4}$

5. $\quad \frac{\boxed{}}{8} = \frac{\boxed{}}{2}$

Complete these fractions.

6. $\dfrac{4}{5} = \dfrac{8}{\boxed{}}$

7. $\dfrac{2}{3} = \dfrac{\boxed{}}{9}$

8. $\dfrac{1}{\boxed{}} = \dfrac{3}{12}$

9. $\dfrac{3}{4} = \dfrac{\boxed{}}{12}$

10. $\dfrac{3}{10} = \dfrac{6}{\boxed{}}$

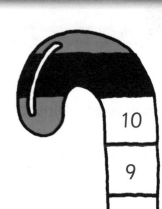

| 10 |
| 9 |
| 8 |
| 7 |
| 6 |
| 5 |
| 4 |
| 3 |
| 2 |
| 1 |

Colour in your score

Test 9 Time

Mornings and **afternoons** are shown by **am** and **pm**.

7.25am ⟶ this is in the morning.

7.25pm ⟶ this is in the evening.

Aston	9.55am	10.45am	11.50am
Banley	10.25am	11.20am	12.25pm
Compton	10.40am	11.40am	12.50pm
Dinsford	11.30am	12.20pm	1.45pm

11.35

hours   minutes
past the
hour

How many minutes do these train journeys take?

1. 9.55am Aston ⟶ Banley [ ] minutes

2. 11.20am Banley ⟶ Compton [ ] minutes

3. 12.50pm Compton ⟶ Dinsford [ ] minutes

4. 10.45am Aston ⟶ Compton [ ] minutes

5. 10.25am Banley ⟶ Dinsford [ ] minutes

Draw the hands on each clock to show the time.

6. **8.55**

7. **3.45**

8. **10.35**

9. **12.40**

10. **1.05**

Colour in your score

40

Test 10 Data handling (1)

This **pictogram** shows information about 4 buses that make the same journey at different times.

Bus	Number of people on each bus
A	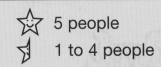
B	☆ ☆
C	☆ ☆ ☆ ☆ ☆ ⚡
D	☆ ☆ ☆

☆ 5 people

⚡ 1 to 4 people

Passengers on bus C

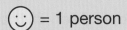

| Adults | Children | Babies |

😊😊 = 2 people   😊 = 1 person

1. How many people travelled on bus B? ☐

2. How many people travelled on bus D? ☐

3. Approximately how many people travelled on bus A?

☐ to ☐ people

4. Approximately how many people travelled on bus C?

☐ to ☐ people

5. Approximately how many people travelled altogether on all 4 buses?

☐ to ☐ people

6. How many adults travelled on bus C? ☐

7. How many children travelled on bus C? ☐

8. How many babies travelled on bus C? ☐

9. How many more adults than children travelled on bus C? ☐

10. How many people travelled altogether on bus C? ☐

Colour in your score

41

Test 11 Multiplying and dividing by 10

To multiply by 10, move all the digits to the **left**. The empty place is filled by a zero.

$$75 \times 10 =$$
$$750$$

To divide by 10, move all the digits one place to the **right**.

$$230 \div 10 =$$
$$23$$

Multiply each of these numbers by 10.

1. | 45 | x 10 |
2. | 63 | x 10 |
3. | 81 | x 10 |
4. | 107 | x 10 |
5. | 234 | x 10 |

Divide each of these numbers by 10.

6. | 530 | ÷10 |
7. | 470 | ÷10 |
8. | 380 | ÷10 |
9. | 6350 | ÷10 |
10. | 8010 | ÷10 |

Colour in your score

42

Test 12 **Addition**

Use mental methods to answer these.

1. 48 + 30 =

2. 36 + 23 =

3. 44 + 46 =

4. 38 + 70 =

5. 56 + 29 =

6. 48 + 37 =

7. 81 + 63 =

8. 72 + 49 =

9. 39 + 45 =

10. 57 + 74 =

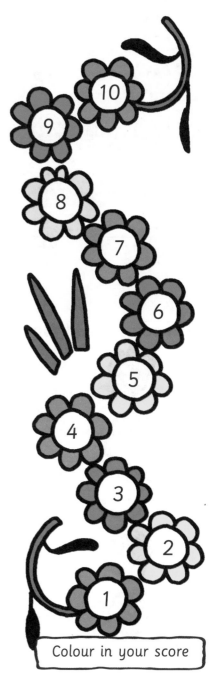

Colour in your score

43

Test 13 Money: adding coins

When adding coins, start with the **highest value** coins to make it easier.

Write these totals.

1. £1   20p   20p   50p   2p ⇨ ☐

2. £1   £1   20p   50p   10p ⇨ ☐

3. 50p   20p   £2   2p   5p ⇨ ☐

4. 10p   1p   2p   £2   £2 ⇨ ☐

5. 10p   50p   2p   5p   £1 ⇨ ☐

Which coins would you use to buy these books?

6. £4.90 _____

7. £3.50 _____

8. £1.13 _____

9. £2.26 _____

10. £4.14 _____

Colour in your score

44

Test 14 Measures problems

When doing **measures problems**, make sure you read the questions carefully and then work out what calculations you need to do.

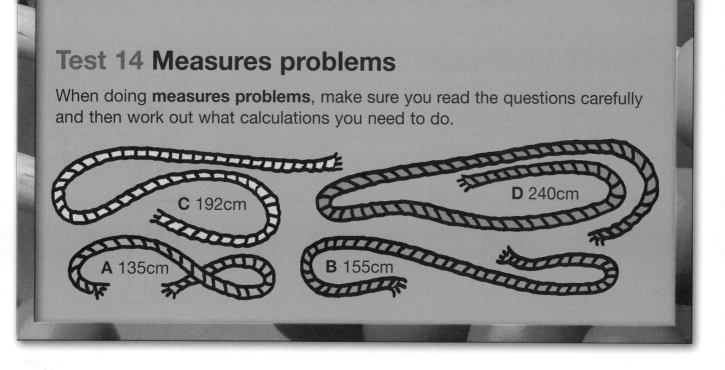

1. What is the difference in length between the longest and shortest ropes? ⬚ cm

2. How much longer is rope C than rope B? ⬚ cm

3. What is the total length of ropes A and B? ⬚ cm

4. Which rope is 85cm longer than rope B? ⬚

5. Which rope can be cut into 5 equal lengths of 27cm? ⬚

Work out the answers to these problems.

6. A chef has a 630g bag of flour and uses 85g. How much flour is left in the bag? ⬚ g

7. Alex drove 5800km in one year and 7600km the following year. How much further did he drive in the second year? ⬚ km

8. If 38g of cake mixture is needed to make 1 cake, how much is needed to make 6 cakes? ⬚ g

9. Vikram swam 850m for a sponsored swim. He swam in widths of 10m. How many widths did he swim? ⬚

10. A pack of 6 cartons has 1260ml of drinks in total. A can holds 240ml. Which holds more, a can or a carton? ⬚

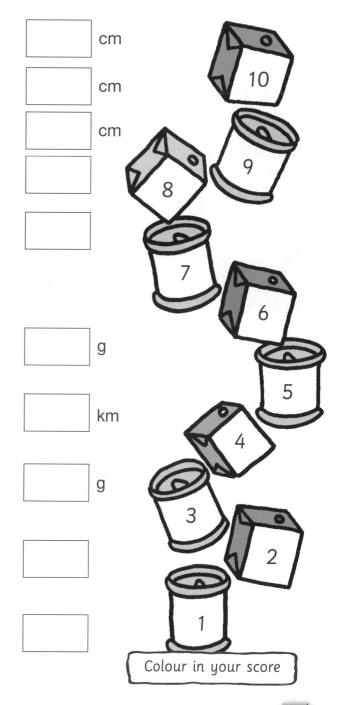

Colour in your score

45

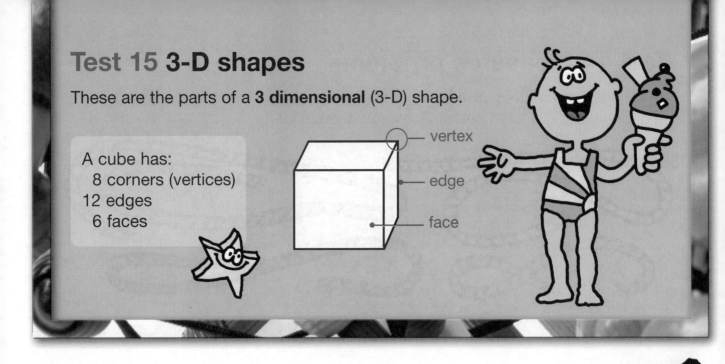

Test 15 3-D shapes

These are the parts of a **3 dimensional** (3-D) shape.

A cube has:
  8 corners (vertices)
12 edges
  6 faces

vertex

edge

face

Name each shape. Write the missing numbers of corners, edges or faces.

1. name _____

2. ☐ corners 8 edges 3. ☐ faces

4. name _____

0 corners 5. ☐ edges 6. ☐ faces

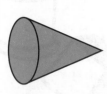

7. name _____

1 corners 1 edges 8. ☐ faces

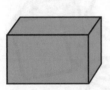

9. name _____

8 corners 12 edges 10. ☐ faces

Colour in your score

46

Test 16 **Number patterns**

Look for number patterns.

1	2	3	4	5	6
7	8	9	10	11	12
13	14	15	16	17	18
19	20	21	22	23	24
25	26	27	28	29	30
31	32	33	34	35	36

Write the next number in each number pattern.

1. | 12 | 14 | 16 | 18 | 20 | |

2. | 15 | 18 | 21 | 24 | 27 | |

3. | 9 | 11 | 13 | 15 | 17 | |

4. | 33 | 30 | 27 | 24 | 21 | |

5. | 16 | 20 | 24 | 28 | 32 | |

Write the missing number in each number pattern.

6. | 22 | 20 | 18 | | 14 | 12 | 10 |

7. | 27 | 24 | | 18 | 15 | 12 | 9 |

8. | | 28 | 30 | 32 | 34 | 36 | 38 |

9. | 36 | 32 | 28 | | 20 | 16 | 12 |

10. | 9 | 12 | 15 | 18 | | 24 | 27 |

Colour in your score

47

Test 17 Division

Use multiplication to help work out **division** questions.

$$24 \div 6 = \boxed{} \Rightarrow \begin{array}{l} 6 \times \boxed{} = 24 \\ 6 \times 4 = 24 \end{array}$$

$$24 \div 6 = 4$$

If a number cannot be divided exactly, it leaves a remainder.

$$26 \div 4 = 6 \text{ remainder } 2$$

Answer these.

1. $30 \div 5 =$

2. $32 \div 4 =$

3. $42 \div 3 =$

4. $52 \div 2 =$

5. $85 \div 5 =$

Answer these and write the remainder.

6. $34 \div 4 =$ _____ remainder _____

7. $29 \div 2 =$ _____ remainder _____

8. $58 \div 5 =$ _____ remainder _____

9. $47 \div 3 =$ _____ remainder _____

10. $86 \div 10 =$ _____ remainder _____

Colour in your score

48

Test 18 Money problems (1)

When finding the **difference** between two amounts, **count on** from the **lower** amount.

The **difference** between £1.80 and £3.30 is **£1.50** (20p + £1 + 30p).

+ 20p   + £1   + 30p

£1.80   £2.00   £3.00   £3.30

Write the difference between these prices.

1. £2.40 £3.50

2. £1.70 £2.25

3. £2.40 £1.60

4. £2.34 £1.50

5. £1.95 £1.10

6. £3.45 £2.90

7. £4.72 £2.80

8. £1.30 £4.65

9. £2.63 £1.20

10. £4.18 £2.80

10
9
8
7
6
5
4
3
2
1

Colour in your score

49

Test 19 Fractions (2)

$\frac{1}{3}$ of 15 is the same as 15 ÷ 3 = 5

Work out the answers.

1. $\frac{1}{4}$ of   12   =

2. $\frac{1}{2}$ of   28   =

3. $\frac{1}{3}$ of   18   =

4. $\frac{1}{5}$ of   20   =

5. $\frac{1}{4}$ of   16   =

6. $\frac{1}{10}$ of   60   =

7. $\frac{1}{3}$ of   24   =

8. $\frac{1}{5}$ of   35   =

9. $\frac{1}{4}$ of   32   =

10. $\frac{1}{10}$ of   90   =

Colour in your score

50

Test 20 Data handling (2)

The numbers 1-10 have been sorted on these two diagrams.

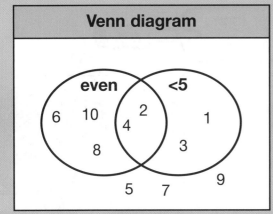

Venn diagram	Carroll diagram

Venn diagram: even, <5 — 6, 10, 8 (even); 4, 2 (overlap); 1, 3 (<5); 5, 7, 9 (outside)

Carroll diagram:
	even	not even
<5	2 4	1 3
not <5	6 10 8	5 9 7

Write the numbers in the correct place on each diagram.

> means greater than
< means less than

1. 7
2. 31
3. 28
4. 16
5. 19

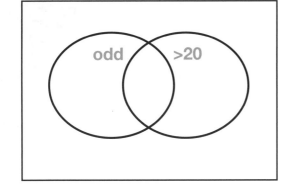

Venn diagram: odd, >20

6. 24
7. 13
8. 15
9. 1
10. 6

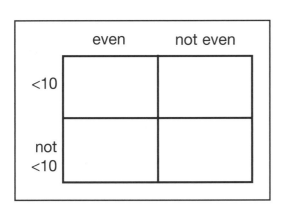

	even	not even
<10		
not <10		

Colour in your score

51

Test 21 Place value (2)

To help work out the **order of numbers**, you can write them in a list, lining up the units columns.

247 29 2403 249

2403
249
247
29

> greater than
< less than

Write the signs > or < for each pair of numbers.

1. 6093 ☐ 6103

2. 4206 ☐ 4311

3. 7415 ☐ 7409

4. 2046 ☐ 2050

5. 8114 ☐ 8108

Write the numbers in order starting with the smallest.

6. 308 318 381 310 ☐ ☐ ☐ ☐

7. 4180 4081 4191 4008 ☐ ☐ ☐ ☐

8. 6293 6095 6120 6905 ☐ ☐ ☐ ☐

9. 2004 2140 410 4010 ☐ ☐ ☐ ☐

10. 902 9214 9189 989 ☐ ☐ ☐ ☐

Colour in your score

52

Test 22 Subtraction

There are lots of ways to **take one number from another**. Look
at the numbers carefully and work out the best mental method to
use for those numbers.

Use mental methods to answer these.

1.   42   –   29   =   [ ]

2.   57   –   23   =   [ ]

3.   31   –   17   =   [ ]

4.   62   –   31   =   [ ]

5.   54   –   19   =   [ ]

6.   81   –   4   =   [ ]

7.   305   –   9   =   [ ]

8.   63   –   38   =   [ ]

9.   52   –   7   =   [ ]

10.   89   –   35   =   [ ]

10
9
8
7
6
5
4
3
2
1

Colour in your score

Test 23 **Area**

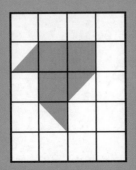 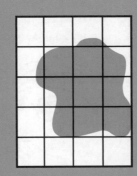

To work out the area of an **irregular shape**, count the whole squares.

$\frac{1}{2}$ or more squares count as whole squares.

Ignore squares less than $\frac{1}{2}$.

For shapes with **straight sides**, count $\frac{1}{2}$ squares.

Work out the areas of these shapes.

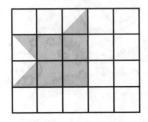 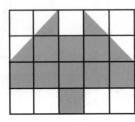 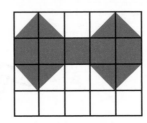

1. ☐ squares   2. ☐ squares   3. ☐ squares

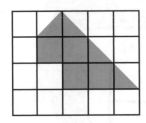 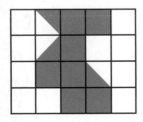 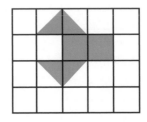

4. ☐ squares   5. ☐ squares   6. ☐ squares

Work out the approximate areas of these shapes.

7.   8.   9.   10.

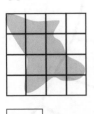   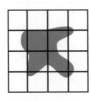

☐ squares   ☐ squares   ☐ squares   ☐ squares

Colour in your score

54

Test 24 **Shape: symmetry**

A shape has line **symmetry** if both sides are exactly the same when a mirror line is drawn.

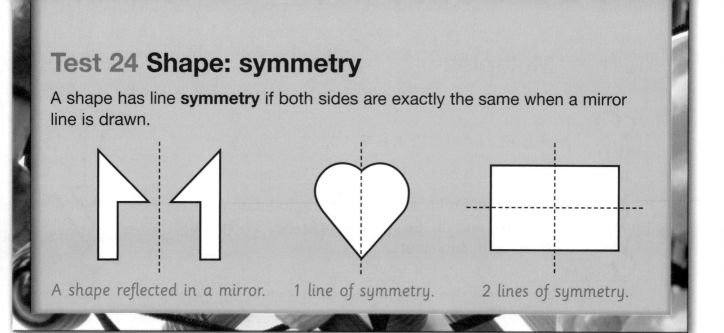

A shape reflected in a mirror.   1 line of symmetry.   2 lines of symmetry.

Draw the reflection of each shape.

1.   2.   3.

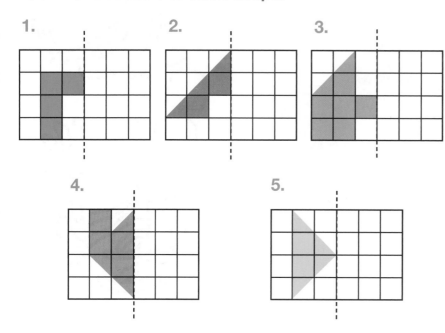

4.   5.

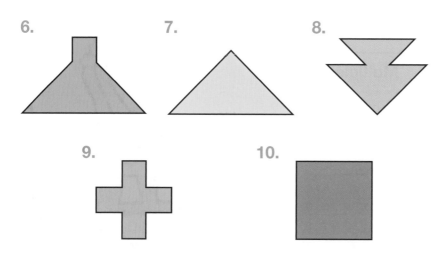

Draw the lines of symmetry on each shape.

6.   7.   8.

9.   10.

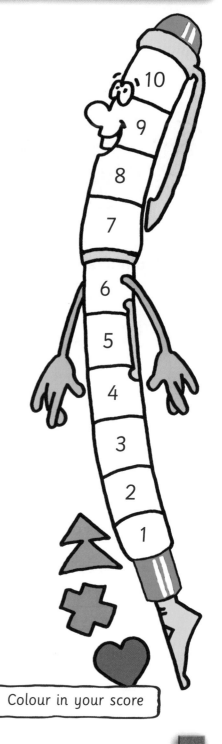

Colour in your score

Test 25 Multiples

Multiples of 2 are: 2, 4, 6, 8, 10, 12... and so on.

Multiples of 3 are: 3, 6, 9, 12, 15, 18... and so on.

Multiples of a number do not come to an end at x10, they go on and on. So, for example, 82, 94, 106 and 300 are all multiples of 2.

Which of these numbers are multiples of 2, 3, 4 or 5?
Some numbers are used more than once.

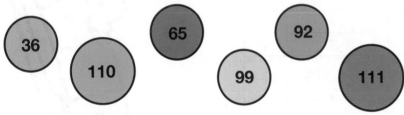

36   110   65   99   92   111

Multiples of 2

1.

2.

3.

Multiples of 3

4.

5.

6.

Multiples of 4

7.

8.

Multiples of 5

9.

10.

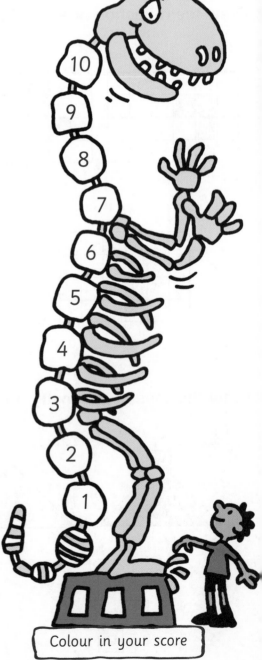

Colour in your score

Test 26 Multiplication

When **multiplying** it can help to break numbers up.

43 × 5 =	

$$40 \times 5 = 200$$
$$3 \times 5 = + 15$$
$$43 \times 5 = 215$$

$$\begin{array}{r} 4\,3 \\ \times \quad 5 \\ \hline 2\,1\,5 \\ 1 \end{array}$$

Answer these.

1. 36 × 3 = ☐

2. 41 × 4 = ☐

3. 53 × 2 = ☐

4. 47 × 3 = ☐

5. 56 × 4 = ☐

Answer these.

6.
$$\begin{array}{r} 5\,3 \\ \times \quad 3 \\ \hline \end{array}$$

7.
$$\begin{array}{r} 8\,4 \\ \times \quad 2 \\ \hline \end{array}$$

8.
$$\begin{array}{r} 6\,7 \\ \times \quad 4 \\ \hline \end{array}$$

9.
$$\begin{array}{r} 7\,4 \\ \times \quad 3 \\ \hline \end{array}$$

10.
$$\begin{array}{r} 5\,9 \\ \times \quad 5 \\ \hline \end{array}$$

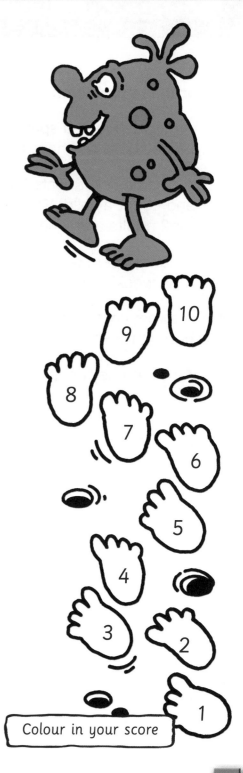

Colour in your score

Test 27 **Money problems (2)**

When working out **word problems**, read the questions carefully to work out the calculations you need to do.

Answer these problems.

1. Amy has 95p and spends 57p. How much money does she have left?

2. A cinema ticket costs £3.50 for an adult and £3 for a child. What is the total cost for 2 adults and 2 children?

3. A newspaper costs 35p. What is the cost for a week's supply of newspapers?

4. A sweet costs 14p. How many can be bought for £1?

5. A bus journey costs £1.20. How much will the total fare be for 4 people?

6. A car costs £2400. If it is reduced by £800, how much will it cost?

7. A book costs £4.70. It is reduced by £1.90 in a sale. What is the new price of the book?

8. Sam has two 20p coins and a 50p coin. He buys a magazine at 72p. How much money does he have left?

9. What is the total cost of a £4.50 T-shirt and a £3.70 pair of shorts?

10. If a fairground ride costs 80p, what is the cost of 3 rides?

Colour in your score

Test 28 **Decimals**

A **decimal point** is used to separate whole numbers from fractions.

$0 . 1 = \frac{1}{10}$ $0 . 2 = \frac{2}{10}$ $0 . 5 = \frac{1}{2}$

tens	units		tenths
8	**2**	**·**	**6**
80	2		$\frac{6}{10}$

Change these fractions to decimals.

1. $\frac{7}{10}$ =

2. $1\frac{1}{2}$ =

3. $3\frac{3}{10}$ =

4. $\frac{9}{10}$ =

5. $2\frac{4}{10}$ =

Write the decimals on this number line.

6. ☐   7. ☐   8. ☐   9. ☐   10. ☐

0 ———————————————————— 1

Test 29 **Time problems**

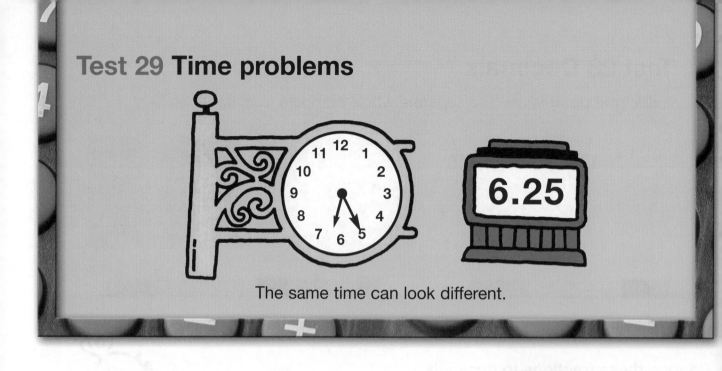

The same time can look different.

These clocks are 45 minutes fast.
Write the real time for each of them.

1.

4.

2.

5.

3.

A train takes 20 minutes between each of these stations.
Complete the timetable.

6.	Smedley	2.10	
7.	Chadwick		4.45
8.	Welby	2.50	
9.	Burnsford		5.25
10.	Ragby		5.45

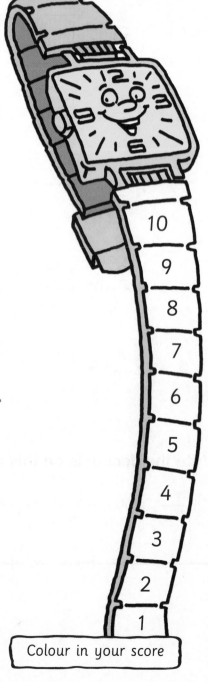

Colour in your score

Test 30 Data handling (3)

These **graphs** show the number of cans collected by two classes in a school over a month.

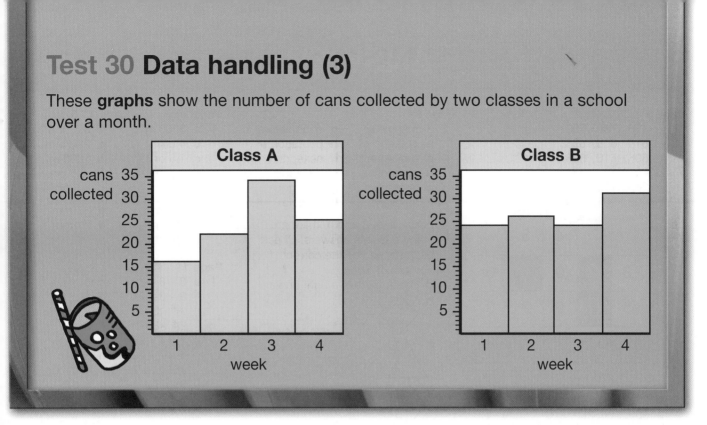

Answer these problems.

1. How many cans were collected by Class A in week 2?

2. How many cans were collected by Class B in week 1?

3. In which week did Class A collect 25 cans?

4. How many more cans were collected in week 1 by Class B than by Class A?

5. In which week did Class A collect 10 more cans than Class B?

6. In which week did Class B collect 26 cans?

7. In which 2 weeks were the same number of cans collected by Class B?

8. How many more cans were collected in week 3 by Class A than by Class B?

9. How many cans altogether were collected by Class B?

10. Which class collected the most cans?

Colour in your score

ANSWERS

Page 2

I
- **a** 758, 759, 760
- **b** 618, 608, 598
- **c** 506, 516, 526
- **d** 741, 841, 941
- **e** 285, 185, 85
- **f** 902, 901, 900

II

	¹7	²4	3	
³3	■	0	■	⁴6
5	■	⁵9	2	0
	1			8

Page 3

I
- **a** 29, 41, 45, 49 — rule +4
- **b** 300, 290, 280, 270 — rule −10
- **c** 144, 140, 136, 132 — rule −2
- **d** 77, 83, 86, 92 — rule +3
- **e** 112, 107, 102, 92 — rule −5
- **f** 40, 46, 58, 76 — rule +6

II
- **a** −4, −3, −1, 1, 2
- **b** −3, −2, −1, 2, 3
- **c** −7, −6, −5, −3, −1, 0
- **d** −2, −1, 0, 1, 2, 4

Page 4

I
- **a** 5 tens
- **b** 6 thousands
- **c** 8 ones
- **d** 2 hundreds
- **e** 6 tens
- **f** 5 thousands
- **g** 9 ones
- **h** 7 hundreds
- **i** 3 tens
- **j** 4 ones
- **k** 8 thousands
- **l** 7 hundreds

II
- **a** 3850
- **b** 7900
- **c** 3680
- **d** 4120
- **e** 9000
- **f** 465
- **g** 291
- **h** 340
- **i** 507
- **j** 800

Page 5

I
- **a** 12, 120, 1200
- **b** 15, 150, 1500
- **c** 15, 150, 1500
- **d** 7, 70, 700
- **e** 8, 80, 800
- **f** 9, 90, 900
- **g** 120
- **h** 1300
- **i** 1300
- **j** 1800
- **k** 240
- **l** 40
- **m** 1600

II

34	51	59	41	76	82	38	62
66	75	25	65	24	47	53	77
91	19	72	83	17	45	96	13
24	81	74	56	35	55	48	52

Page 6

I
- **a** triangle ✔
- **b** octagon ✔
- **c** pentagon
- **d** octagon
- **e** heptagon
- **f** hexagon
- **g** quadrilateral
- **h** pentagon ✔
- **i** hexagon ✔
- **j** decagon
- **k** nonagon
- **l** triangle

II For a, b, c, d and h, there are many possible answers. Check each shape has the correct number of sides.
- **d** check there is a right angle.
- **e**
- **f**
- **g**

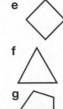

Page 7

I
- **a** £958, £1090, £1900, £2589, £2850
- **b** 965 km, 2830 km, 3095 km, 3520 km, 3755 km
- **c** 1599 g, 1995 g, 2046 g, 2460 g, 2604 g
- **d** 4599 ml, 4600 ml, 7025 ml, 7028 ml, 7529 ml

II 2389, 2398, 2839, 2893, 2938, 2983
3289, 3298, 3829, 3892, 3928, 3982
8239, 8293, 8329, 8392, 8923, 8932
9238, 9283, 9328, 9382, 9823, 9832

Page 8

I
- **a** 4.12
- **b** 10.38
- **c** 12.26
- **d** 5.34
- **e**
- **f**
- **g**
- **h**

II
- **a** 35 minutes
- **b** 40 minutes
- **c** 40 minutes
- **d** 55 minutes

Page 9

I
- **a** 6, 3, 4, 2
- **b** 10, 5, 4, 2
- **c** 12, 8, 6, 4

II
- **a** any 18 squares red, any 9 blue, any 6 yellow. 3 left white.
- **b** any 24 squares red, any 12 blue, any 8 yellow. 4 left white.

Page 10

I
- **a** 3500 m
- **b** 4 cm
- **c** 1.5 or $1\frac{1}{2}$ m
- **d** 80 mm
- **e** 25 cm
- **f** 6.5 or $6\frac{1}{2}$ km
- **g** 220 mm
- **h** 1800 m
- **i** 475 cm
- **j** 6.5 or $6\frac{1}{2}$ cm

II
- **a** 45mm
- **b** 62 mm
- **c** 58 mm
- **d** 37 mm
- **e** 71 mm

Page 11

I
- **a** 3 × 8 = 24
  8 × 3 = 24
  24 ÷ 3 = 8
  24 ÷ 8 = 3
- **b** 7 × 4 = 28
  4 × 7 = 28
  28 ÷ 4 = 7
  28 ÷ 7 = 4
- **c** 6
- **d** 4
- **e** 21
- **f** 5
- **g** 6
- **h** 24
- **i** 4
- **j** 8

II
- 60 ÷ 9 → 6
- 93 ÷ 10 → 3
- 89 ÷ 5 → 4
- 38 ÷ 6 → 2
- 106 ÷ 10 → 6
- 61 ÷ 2 → 1
- 37 ÷ 3 → 1
- 48 ÷ 5 → 3
- 65 ÷ 6 → 5
- 80 ÷ 3 → 2
- 53 ÷ 6 → 5
- 46 ÷ 6 → 4

Page 12

I
- **a** >
- **b** <
- **c** >
- **d** <
- **e** >
- **f** >
- **g** >
- **h** <
- **i** <
- **j** >
- **k** <
- **l** >

II
- **a** 4168, 4167, 4166, 4165
- **b** 3839, 3840, 3841,
- **c** 9001, 9000, 8999, 8998, 8997
- **d** 4422, 4423, 4424, 4425
- **e** 7081, 7080, 7079, 7078, 7077

Page 13

I
- **a** cuboid
- **b** cylinder
- **c** cone
- **d** sphere
- **e** cube
- **f** pyramid

II

	faces	edges	vertices
a	5	8	5
b	6	12	8
c	5	9	6
d	4	6	4

Page 14

I
- **a** 2 kg
- **b** 1500 g
- **c** 5.5 or $5\frac{1}{2}$ kg
- **d** 1.25 or $1\frac{1}{4}$ kg
- **e** 7000 g
- **f** 3250 g
- **g** 10 000 g
- **h** 6.75 or $6\frac{3}{4}$ kg
- **i** 2500 g
- **j** 4750 g
- **k** 9.5 or $9\frac{1}{2}$ kg
- **l** 1750 g

II
- **a** 2.5 or $2\frac{1}{2}$ kg
- **b** 8 kg
- **c** 6.5 or $6\frac{1}{2}$ kg
- **d** 550 g
- **e** 800 g
- **f** 300 g

Page 15

I
- **a** 94
- **b** 101
- **c** 64
- **d** 148
- **e** 165
- **f** 126
- **g** 131
- **h** 141
- **i** 171
- **j** 256
- **k** 251
- **l** 301

II **a** 444   **d** 1001
   b 355   **e** 754
   c 620   **f** 810

Page 16
I **a** 10 squares   **d** 33 squares
   b 21 squares   **e** 23 squares
   c 13 squares

II Check the shapes drawn use 8 squares.

Page 17
I **a** 16 cm   **d** 20 cm
   b 14 cm   **e** 14 cm
   c 20 cm

II **a** 14 cm   **c** 10 cm
   b 14 cm   **d** 12 cm

Page 18
I **a** ... **d** ...
   b ... **e** ...
   c ... **f** ...

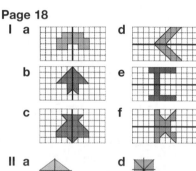

II **a** 2 lines of symmetry   **d** 4 lines of symmetry
   b 3 lines of symmetry   **e** 5 lines of symmetry
   c 6 lines of symmetry

Page 19
I **a** 3 l   **g** 3500 ml
   b 1500 ml   **h** 2 l
   c 6000 ml   **i** 4.5 or $4\frac{1}{2}$ l
   d 2.25 or $2\frac{1}{4}$ l   **j** 8750 ml
   e 1.75 or $1\frac{3}{4}$ l   **k** 5.75 or $5\frac{3}{4}$ l
   f 10 000 ml   **l** 6750 ml

II **a** 500 ml   **e** 1000 ml
   b 750 ml   **f** 300 ml
   c 100 ml   **g** 500 ml
   d 250 ml   **h** 1500 ml

Page 20
I **a** 0.3   **e** 1.7   **i** 42.5
   b 0.5   **f** 3.3   **j** 19.4
   c 0.2   **g** 4.5   **k** 68.8
   d 0.7   **h** 7.9   **l** 59.6

II **a** 0.2, 0.4, 0.6, 0.7, 0.9
   b 3.1, 3.4, 3.6, 3.7, 3.9
   c 0.4, 0.8, 1.2, 1.6

Page 21
I **a** 25 jumps
   b Ashley
   c Alex
   d 20 more
   e Kim
   f 7 more
   g Jo – 32 skips
     Sam – 25 skips
     Alex – 37 skips
     Kim – 20 skips
     Ashley – 45 skips

II Check pictogram is accurate.

Page 22
I **a** 36   **c** 36   **e** 76
   b 35   **d** 38   **f** 69

II **a** 26 kg   **c** 60 kg   **e** 76 kg
   b 28 kg   **d** 46 kg   **f** 77 kg

Page 23
I **a** $\frac{4}{6} = \frac{2}{3}$ **f** 3
   b $\frac{6}{10} = \frac{3}{5}$ **g** 5
   c $\frac{4}{8} = \frac{1}{2}$ **h** 5
   d $\frac{6}{8} = \frac{3}{4}$ **i** 12
   e 4   **j** 3

II **a** $\frac{7}{15}$ **b** $\frac{5}{25}$ **c** $\frac{8}{27}$

Page 24
I

Total drop (m)	Nearest 10m	Nearest 100m
979	980	1000
947	950	900
774	770	800
739	740	700
646	650	600
581	580	600
561	560	600

II **a** 160   **c** 1000   **e** 200
   b 200   **d** 1150   **f** 210

Page 25
I Multiples of 2 – 48, 56, 100, 86, 52, 82, 42, 70, 60
Multiples of 3 – 48, 39, 42, 63, 60
Multiples of 4 – 48, 56, 100, 52, 60
Multiples of 5 – 100, 85, 70, 115, 60, 65

II You can see these patterns:
The red squares make diagonal lines and the blue squares make two vertical lines.
The ones digits in the numbers on each blue line are the same: 5s or 0s.
Starting at the top of each red line, the tens digits ascend, while the ones digits descend, e.g. the first line is 3, 12, 21: tens digits (0),1, 2; ones digits 3, 2, 1.

Page 26
I **a** £1.75   **c** £1.44   **e** 28p
   b £1.35   **d** £1.45   **f** 76p

II **a** £6.51   **e** £2.41
   b £1.01   **f** £1.11
   c £2.11   **g** £2.61
   d £6.31   **h** £3.41

Page 27
I **a**

greater
   b

smaller
   c

smaller
   d

greater

II **a** East   **d** East
   b East   **e** South-east
   c South-west   **f** West

Page 28
I **a**
 →
   b 6 : 20 → 8 : 40
   c

   d 1 : 45 → 3 : 15
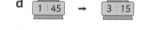

II **a** 15 minutes   **d** 15 minutes
   b 25 minutes   **e** 20 minutes
   c 5 minutes   **f** 25 minutes

Page 29
I **a** silver   **c** 10   **e** 9
   b 12   **d** blue   **f** 106

II **a** 45   **c** 6
   b Thursday   **d** Wednesday

Page 30
I **a** £1.80   **d** 7   **g** 240 km
   b 55 g   **e** 10   **h** 15
   c £5.90   **f** 9

II 150 g margarine
120 g sugar
180 g flour
3 eggs
45 g cocoa powder
60 ml milk

Page 31
I **a** C, R, N
   b D (4,9)   A (7,6)   S (2,0)
   c RECTANGLE

II

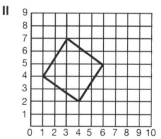

a The shape is a square.
b Check the new coordinates for the rectangle.

Page 32
1. 70
2. 8000
3. 600
4. 90
5. 6000
6. 2108
7. 4090
8. 7235
9. 3816
10. 9700

Page 33
1. 110
2. 60
3. 80
4. 1400
5. 500
6. 1700
7. 1600
8. 90
9. 70
10. 1300

Page 34
1. 50cm
2. 5mm
3. 100m
4. 250ml
5. 250g
6. 38mm
7. 52mm
8. 63mm
9. 26mm
10. 77mm

Page 35
1. 4
2. 8
3. 6
4. 3
5. 5
6. pentagon
7. hexagon
8. octagon
9. quadrilateral
10. triangle

Page 36
The missing numbers are in **bold**.
1. 32 35 38 **41** 44 47 50 53 **56** 59
2. 48 52 **56** 60 64 68 **72** 76 80 84
3. 31 29 27 **25** 23 21 **19** 17 15
4. 230 210 **190** 170 150 **130** 110 90 70
5. 76 81 86 91 **96** 101 106 **111** 116
6. − 2
7. 4
8. − 7
9. − 3
10. − 1

Page 37
1. 5
2. 10
3. 24
4. 4
5. 9
6. 5
7. 60
8. 9
9. 6
10. 9

Page 38
1. 235p
2. 645p
3. £3.70
4. 109p
5. £2.14
6. 275p
7. £2.55
8. £3.15
9. £4.05
10. £4.10

Page 39
1. $\frac{4}{10}$ = $\frac{2}{5}$
2. $\frac{3}{6}$ = $\frac{1}{2}$
3. $\frac{2}{8}$ = $\frac{1}{4}$
4. $\frac{6}{8}$ = $\frac{3}{4}$
5. $\frac{4}{8}$ = $\frac{1}{2}$

6. $\frac{8}{10}$
7. $\frac{6}{9}$
8. $\frac{1}{4}$
9. $\frac{9}{12}$
10. $\frac{6}{20}$

Page 40
1. 30 minutes
2. 20 minutes
3. 55 minutes
4. 55 minutes
5. 65 minutes
6.
7.
8.
9.
10.

Page 41
1. 10
2. 15
3. 11 to 14 people.
4. 26 to 29 people.
5. 62 to 68 people.
6. 14
7. 9
8. 5
9. 5
10. 28

Page 42
1. 450
2. 630
3. 810
4. 1070
5. 2340
6. 53
7. 47
8. 38
9. 635
10. 801

Page 43
1. 78
2. 59
3. 90
4. 108
5. 85
6. 85
7. 144
8. 121
9. 84
10. 131

Page 44
1. £1.92
2. £2.80
3. £2.77
4. £4.13
5. £1.67
6. £2 £2 50p 20p 20p
7. £2 £1 50p
8. £1 10p 2p 1p
9. £2 20p 5p 1p
10. £2 £2 10p 2p 2p

Page 45
1. 105cm
2. 37cm
3. 290cm
4. D
5. A
6. 545g
7. 1800km
8. 228g
9. 85
10. can

Page 46
1. pyramid
2. 5 corners
3. 5 faces
4. cylinder
5. 2 edges
6. 3 faces
7. cone
8. 2 faces
9. cuboid
10. 6 faces